Sponsored Technology Workshops 2017

Core Programming Area at the 2017 AIChE Annual Meeting

Minneapolis, Minnesota, USA
29 October - 3 November 2017

ISBN: 978-1-5108-5815-2

Printed from e-media with permission by:

Curran Associates, Inc.
57 Morehouse Lane
Red Hook, NY 12571

Some format issues inherent in the e-media version may also appear in this print version.

Printed by Curran Associates, Inc. (2018)

AIChE
120 Wall Street, FL 23
New York, NY 10005-4020

Phone: (800) 242-4363
Fax: (203) 775-5177

www.aiche.org

Additional copies of this publication are available from:

Curran Associates, Inc.
57 Morehouse Lane
Red Hook, NY 12571 USA
Phone: 845-758-0400
Fax: 845-758-2633
Email: curran@proceedings.com
Web: www.proceedings.com

TABLE OF CONTENTS

Integrating Global Simulation Competition into Senior Design 1
N/A
Simulation and analysis software for Chemical and Process Engineering 2
N/A
Integrating Global Simulation Competition into Senior Design (2) 3
N/A
Integrating Global Simulation Competition into Senior Design (3) 4
N/A
Hands-on workshop: Mechanistic models for the digital design of robust formulated products and their manufacturing processes 5
N/A
Advanced Simulation to Solve & Optimize Process Development Problems 6
N/A
Progress and Challenges in the Structural Characterization of Nanoporous Materials by Physical Adsorption 7
N/A
High Pressure Adsorption Techniques for Assessing Adsorbents for Capture and Storage Applications 8
N/A
Use of Dynamic Breakthrough Measurements for Characterization of Adsorbents and Optimization of Separation Processes 9
N/A
Use of an Advanced Simulation Package for Predicting Breakthrough Behavior of Gas Mixtures from Single-Component Isotherms 10
N/A

Some abstracts in this book refer to images that were not made available by the publisher

510384 Integrating Global Simulation Competition into Senior Design

Abstract:

University of New Hampshire Assistant Professor, Jeffrey Halpern, engaged his students in the 2017 Process Simulation Cup as part of their design course. Dr. Halpern will join Chemstations at this workshop to share learnings from this experience. The Process Simulation Cup is a global competition to optimize a given flowsheet representing a typical optimization problem. The Process Simulation Cup is open to university students, providing a way for them to both test and demonstrate advanced use of a process simulator. Learn more about the Process Simulation Cup below or at https://www.process-simulation-cup.com/about-cup/.

The Process Simulation Cup is a unique chance for chemical engineering students to learn about process simulation and to gain valuable industry exposure. The task is to optimize a given flowsheet representing a typical optimization problem of the chemical, pharmaceutical or life science industries. The challenge problem is believed to have several local optima and a global optimum, but no closed solution. Finding an optimum requires tenacity and ingenuity - both are desirable professional skills for an excellent engineer. Awards are presented for the participant & institution who has submitted the best design variables. In February 2018, the 2017 Process Simulation Cup champion will be announced. The Process Simulation Cup is organized by Chemstations Europe GmbH, Berlin.

510385 Simulation and analysis software for Chemical and Process Engineering

Abstract:

Engineering problems and projects are now cross functional and multi-disciplinary. The industry investments are increasingly more complex driven by for requirements for sustainability, energy efficiency, higher performance, and global competition. Built on a set of well-establish physics based modeling and computational techniques engineering simulation has moved from R&D center use to a broader enterprise deployment.

This workshop is designed to highlight the advancements in computational physics tools with focus on material, Chemical, petrochemical, and pharmaceutical industries. Technical examples will include modeling capabilities and applications in reaction and combustion, multiphase, fouling, erosion- corrosion, mixing, separation. The content and presentation delivered by industry experts are pulled together to benefit current uses of engineering simulation software (CFD, DEM, FEA, Electromagnetic, System) as well as group leaders, managers, professors, and graduate students interested to learn about the latest advancements in physics based simulation software.

510386 Integrating Global Simulation Competition into Senior Design (2)

Abstract:

University of New Hampshire Assistant Professor, Jeffrey Halpern, engaged his students in the 2017 Process Simulation Cup as part of their design course. Dr. Halpern will join Chemstations at this workshop to share learnings from this experience. The Process Simulation Cup is a global competition to optimize a given flowsheet representing a typical optimization problem. The Process Simulation Cup is open to university students, providing a way for them to both test and demonstrate advanced use of a process simulator. Learn more about the Process Simulation Cup below or at https://www.process-simulation-cup.com/about-cup/.

The Process Simulation Cup is a unique chance for chemical engineering students to learn about process simulation and to gain valuable industry exposure. The task is to optimize a given flowsheet representing a typical optimization problem of the chemical, pharmaceutical or life science industries. The challenge problem is believed to have several local optima and a global optimum, but no closed solution. Finding an optimum requires tenacity and ingenuity - both are desirable professional skills for an excellent engineer. Awards are presented for the participant & institution who has submitted the best design variables. In February 2018, the 2017 Process Simulation Cup champion will be announced. The Process Simulation Cup is organized by Chemstations Europe GmbH, Berlin.

510387 Integrating Global Simulation Competition into Senior Design (3)

Abstract:

University of New Hampshire Assistant Professor, Jeffrey Halpern, engaged his students in the 2017 Process Simulation Cup as part of their design course. Dr. Halpern will join Chemstations at this workshop to share learnings from this experience. The Process Simulation Cup is a global competition to optimize a given flowsheet representing a typical optimization problem. The Process Simulation Cup is open to university students, providing a way for them to both test and demonstrate advanced use of a process simulator. Learn more about the Process Simulation Cup below or at https://www.process-simulation-cup.com/about-cup/.

The Process Simulation Cup is a unique chance for chemical engineering students to learn about process simulation and to gain valuable industry exposure. The task is to optimize a given flowsheet representing a typical optimization problem of the chemical, pharmaceutical or life science industries. The challenge problem is believed to have several local optima and a global optimum, but no closed solution. Finding an optimum requires tenacity and ingenuity - both are desirable professional skills for an excellent engineer. Awards are presented for the participant & institution who has submitted the best design variables. In February 2018, the 2017 Process Simulation Cup champion will be announced. The Process Simulation Cup is organized by Chemstations Europe GmbH, Berlin.

510388 Hands-on workshop: Mechanistic models for the digital design of robust formulated products and their manufacturing processes

Abstract:

If you are interested in attending this workshop, please contact Kristen Hunter at k.hunter@psenterprise.com before the meeting or register here: https://www.psenterprise.com/events/americas/2017/workshop/aiche-2017

Join PSE and industry experts for a hands-on software demonstration of the newly released gPROMS Formulated Products, an innovative platform for the digital design of formulated products and their manufacturing processes. Currently, scientists and engineers in the pharmaceutical, biopharmaceutical, food and consumer goods industries face enormous challenges in efficiently bringing products to market with robust manufacturing processes to produce the desired end-use attributes. Historically, upstream process design choices and formulation decisions have been very difficult to tie to final product characteristics.

gPROMS FormulatedProducts allows scientists and engineers to screen formulations for end-user attributes, determine whether they can be manufactured efficiently and robustly, and explore the design space of the whole formulation and manufacturing chain.

510389 Advanced Simulation to Solve & Optimize Process Development Problems

Abstract:

A common challenge in process industries, which comprise of food, pharmaceutical, materials and biotech industries is to bring new products to market in a time and cost effective manner. Mixing is one of the most important unit operations in the production. Poor mixing can cause losses of millions of Dollars/Euros the chemical. Understanding of transport processes (fluid flow, heat transfer & mass transfer)is key for design, troubleshooting and exploring optimum operating conditions.

Computational fluid dynamics (CFD) and particle modeling with Discrete Element Method (DEM) have been identified as key enabling technologies in finding solutions to many of the challenges that surround scale-up; capable of reducing operating costs across manufacturing and quality divisions.

What will we discuss: Fluid flow and hydrodynamic behavior in stirred vessels, Dispersing gas and solids in liquids, Behavior of high viscosity fluids, Fluid structure interactions in mixing vessels etc., Optimization of process equipment.

Who Should Attend?: This workshop is also open to all professors and students who would like to learn how they can use CFD in their senior projects or research projects.

If you are a chemical process engineer, manager who wants to optimize current processes in ways it was never possible before, this event will help you understand the applicability of deploying simulation to achieve these objectives. If you are already using CFD in your engineering work flow, this session will demonstrate that it is not a single point simulation but rather the ability to rapidly and intelligently explore a host designs and processes that brings the most value from simulation.

510377 Progress and Challenges in the Structural Characterization of Nanoporous Materials by Physical Adsorption

Abstract:

Assessing adsorption properties of nanoporous materials (e.g. for gas storage applications) and their structural characterization is crucial for advancing the application of nanoporous materials such as mesoporous zeolites in catalysis, separations, and other industrial processes. Gas adsorption is well suited for this task because it assesses a wide range of pore sizes, spanning the entire micro- and mesopore range. Within the last two decades major progress has been achieved in the understanding of the adsorption and phase behaviour of fluids in ordered nanoporous materials and in the development of advanced approaches based on statistical mechanics such as molecular simulation and density functional theory of inhomogeneous fluids (see reviews [1,2] and references therein). This progress, coupled with the availability of high resolution experimental procedures for the adsorption of various subcritical fluids, has led to major advances in the structural characterization by physical adsorption. These advances and corresponding recommendations for physisorption characterization are also summarized in a new IUPAC technical report which has been very recently published (in 2015, ref. [3]).

Within this context we will discuss in important fundamental aspects of the underlying adsorption mechanisms of fluids in nanoporous materials (e.g., micro-mesoporous materials with hierarchical pore structure [4]) and their significance for physical adsorption characterization

References

[1] M. Thommes,., K.A. Cychosz, . Adsorption 20, 233 (2014).

[2] J. Landers, G.Y., Gor, A.V. Neimark, A.V. Colloids Surfaces A 437, 3 (2013).

[3] M. Thommes, K. Kaneko, A.V. Neimark, J.P. Olivier, F. Rodriguez Reinoso, J. Rouquerol and K.S.W Sing, Pure Appl. Chem. 87, 1051 (2015)

[4] K. A. Cychosz, R. Guillet-Nicolas, J. Garcia-Martinez, J., M. Thommes, M. Chem. Soc. Rev. 46, 389, (2017)

510378 High Pressure Adsorption Techniques for Assessing Adsorbents for Capture and Storage Applications

Abstract:

Use of porous materials for the capture and storage of gasses has been of interest for many years. Removal of CO_2 and other greenhouse gasses from industrial effluent is an important process for reducing the amount of these greenhouse gasses released into the atmosphere. Storage of CO_2 for use in subsequent processes or for sequestration is just as important as its removal. Storage of H_2 and CH_4, among other gasses is important in the transport and use of alternative fuels. Removal of CO_2 from CH_4 or H_2 is another application where adsorbents play a critical role. We will discuss the use of a commercially available high pressure gas sorption apparatus for the characterization of materials for gas capture and storage.

510381 Use of Dynamic Breakthrough Measurements for Characterization of Adsorbents and Optimization of Separation Processes

Abstract:

Use of adsorbents, both naturally occurring and synthetic, for the separation and purification of gasses has been around for many years. Although these processes are well developed, the introduction of new adsorbents and the need to optimize the existing processes, results in the need for better understanding of the competitive adsorption and equilibrium conditions of the gas-solid system. Optimizing the parameters of the separation process in a production environment, even in a pilot plant, can be very costly. The ability to model these systems in a benchtop system, in the lab, under controlled conditions, is a valuable tool in improving the efficiency of these processes. We will discuss the use of a commercially available dynamic breakthrough apparatus for the development and optimization of gas separation and purification processes.

510382 Use of an Advanced Simulation Package for Predicting Breakthrough Behavior of Gas Mixtures from Single-Component Isotherms

Abstract:

The ability to evaluate parameters of a separation process and predict multi-component adsorption of complex systems is a valuable tool in design and optimization of separation and purification systems. Trial and error, using a breakthrough apparatus can be very time consuming and costly, and my not result in a true optimization of the process. We will discuss the use of a powerful dynamic simulation program, which can be used to optimize parameters and predict the competitive adsorption behavior of gas mixtures on various adsorbents.